SILLY SPACE SCIENCE

by Robin Twiddy

designed & illustrated by Amy Li

LERNER PUBLICATIONS ◆ MINNEAPOLIS

First American edition published in 2021 by Lerner Publishing Group, Inc.

Editor: John Wood
Design: Amy Li

Lerner Publications Company
An imprint of Lerner Publishing Group, Inc.
241 First Avenue North
Minneapolis, MN 55401 USA

For reading levels and more information, look up this title at www.lernerbooks.com.

Main body text set in Mikado. Typeface provided by HVD.

Library of Congress Cataloging-in-Publication Data

Names: Twiddy, Robin, author. | Li, Amy, illustrator.
Title: Silly space science / written by Robin Twiddy; designed & illustrated by Amy Li.
Description: First American edition. | Minneapolis, MN: Lerner Publications, 2021. | Series: Silly science | Includes bibliographical references and index. | Audience: Ages 8-12 | Audience: Grades 4-6 | Summary: "Ever wonder what it'd be like in space? It's sillier than you'd think! Stretched spines, poop on the moon, and planets that rain diamonds are just a few things readers will learn about"– Provided by publisher.
Identifiers: LCCN 2020048876 (print) | LCCN 2020048877 (ebook) | ISBN 9781728416502 (library binding) | ISBN 9781728423913 (paperback) | ISBN 9781728418797 (ebook)
Subjects: LCSH: Space sciences—Juvenile literature. | Outer space—Juvenile literature.
Classification: LCC QB500.22 . T95 2021 (print) | LCC QB500.22 (ebook) | DDC 500.5—dc23

LC record available at https://lccn.loc.gov/2020048876
LC ebook record available at https://lccn.loc.gov/2020048877

Manufactured in the United States of America
1-48897-49211-2/24/2021

CONTENTS

Words that look like **THIS** are explained in the glossary on page 31.

SCIENCE: Why So Serious?

We all know science, don't we? Boring people in boring white coats, writing boring things on boring clipboards. Boring, boring, boring, right? Wrong! What if I told you that science can be sillier than you have ever imagined?

Silliness in SPACE

Space. Now that is serious business. Expensive rockets, daring astronauts, and piles of poo on the moon . . . Wait, that doesn't sound very serious, does it? It turns out that space is sillier than anybody thought.

Space is a very silly place!

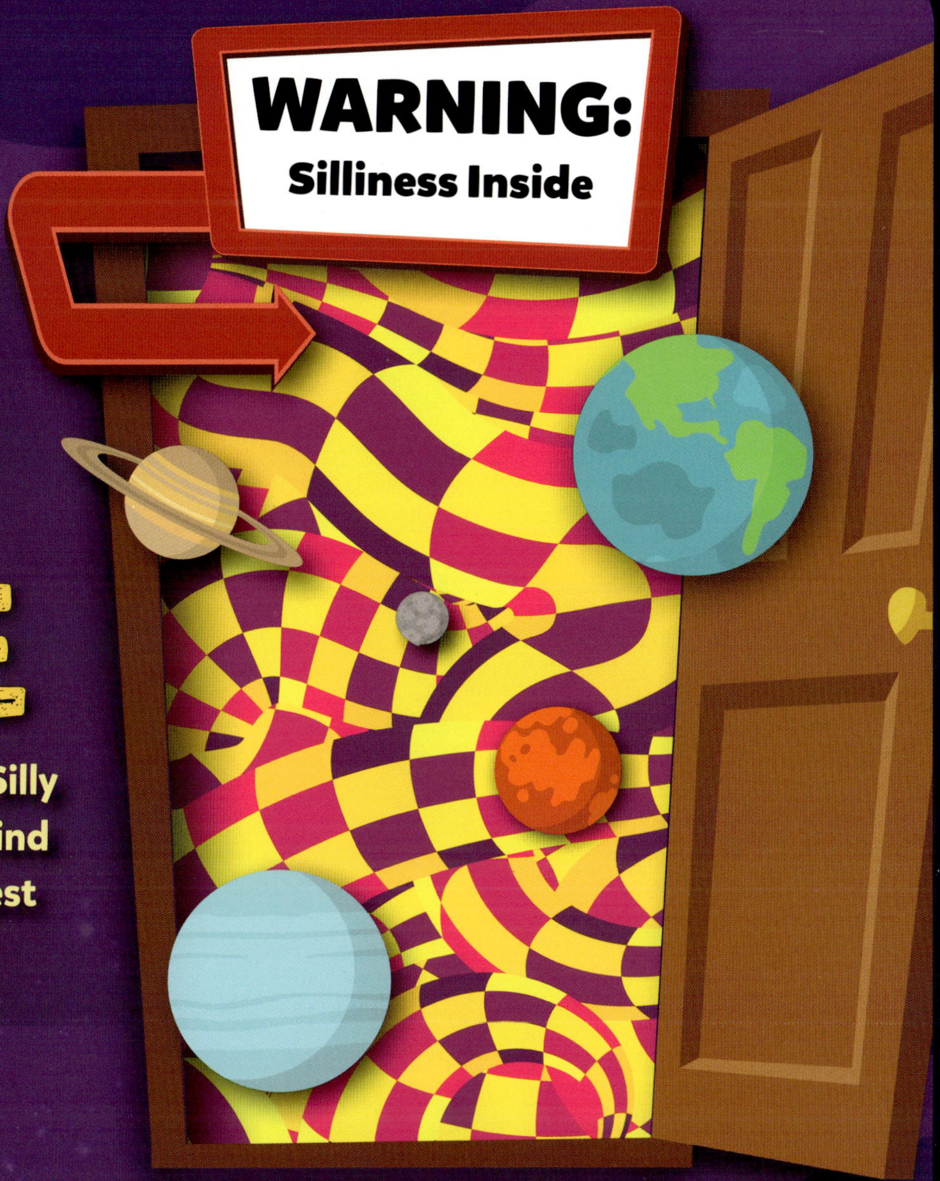

Welcome to the SILLY ZONE

You are about to enter the Silly Zone. In this book you will find scientists doing the strangest things. Get ready!

Real SCIENCE

Even though science can sometimes seem very silly, there is always a very good reason for the silliness. Did you know that there are mirrors on the moon? They are not for checking your reflection when you land. No, they are for bouncing lights from **LASERS** around! Scientists measure how far away the moon is from Earth by shining lasers at the mirrors and seeing how long it takes for the light to bounce back.

PUFFY Heads and Skinny Legs

These astronauts have no idea what they're in for!

WELCOME to the Unknown

Let's be honest, it was pretty silly going into space in the first place, but humans did it. Even though scientists did lots of tests, nobody could be completely sure what would happen to the astronauts' bodies in space.

The International Space Station (ISS) is a joint project between five different SPACE AGENCIES.

BODIES in Space

Being in space for a few hours won't have much of an effect on the human body. However, if you stayed on the ISS for a while, you might notice some pretty strange things happening to your body.

Calling

GROUND CONTROL...

My Legs Feel Funny

After a few weeks of being in space, something odd happens. All the liquid in the body is **REDISTRIBUTED**. A lot more liquid finds its way to the top of your body than it would on Earth. And this means you get a puffy head and skinny legs.

HOW

Does This Happen?

On Earth, your heart works really hard against **GRAVITY** to pump blood all around your body. So, if you take gravity away, what happens? Your heart pumps too hard and sends most of the blood to the head. This leaves the legs **DEPRIVED** of blood and a little skinnier and sillier than usual.

Astronauts sometimes get very dizzy back on Earth. You could say that the gravity goes to their heads.

TALLER Than Before

You Must Be THIS Tall to Ride

Have you ever been to a fair and seen those signs that say, “you must be this tall to ride,” but you were not “this tall”? What if there was a way to become a little bit taller? Would you do it? What if I told you that you would need to train for years, become an astronaut, travel to space, stay there for a few weeks, and then go to the fair as soon as you get back?

What on (or off) Earth is stretching astronauts?

TALL Astronauts

Astronauts have been reaching things on the top shelves for years. In space, they just float to the top shelf. And when they come back to Earth, they are 3 PERCENT taller than they were when they left.

You Are Taller in the MORNING

This doesn't just happen to astronauts—it happens to everyone. When you get up in the morning, you are a bit taller than when you went to bed. You aren't growing in the night—you are actually shrinking in the day. Gravity is pulling your body down all day long. This leads to your spine becoming pushed together. By the end of the day, you are a little shorter.

In Space, It Is ALWAYS Morning

The same thing that happens to your spine during the night is happening to astronauts while they are in space. Without gravity pulling down on them, the astronauts' spines stretch out to their full length.

Space TOILETS: The Final Frontier

Are You AFRAID to Poo in Space?

Are you afraid to poo in space? You should be. Did you know that gravity plays a very big part when you go for a poo? Well, it does. Not only does it help us to push, but it also helps make sure that the poo stays where it's supposed to be—in the toilet.

Read on to find out what happened to all those bags of poo!

Before the Toilet, There Was the BAG!

Before the space toilet was invented, astronauts had to poo in little bags. This wouldn't be easy on Earth, and without gravity's help this was even harder in space. More than one astronaut pooed themselves in space.

The Rogue POO of Apollo 10

Something went wrong on the **APOLLO 10 MISSION**. When the astronauts were returning from the moon, one astronaut noticed a UFO (unidentified flying object) inside the spacecraft. It wasn't a flying saucer—it was a floating poo! One of the crew, Tom Stafford, said, "Give me a napkin, quick. There's a poo floating through the air." The rest of the crew quickly decided to let Stafford deal with the poo.

The space toilet has an On button instead of a flush!

The VACUUM Toilet

Scientists are very good at solving problems—even silly problems such as floating poos in space. Their idea for a space toilet was very similar to a vacuum cleaner. That's right—space toilets suck the waste straight out of the astronauts.

POO on the Moon

WHERE, oh Where Does All the Poo Go?

Did you know that some astronaut poo was fired off into space? Think about that next time you wish on a shooting star. Some poo was left on the moon. Just think—up there on the moon with Neil Armstrong's famous footprint and the US flag are bags and bags of people's poo.

Lunar SCOOPER

Scientists are planning a trip to the moon to collect the bags of poo left by the Apollo 11 team. It is not just poo—there are bags of urine and vomit too. But why are we going back to the moon to collect it?

ACCIDENTAL Experiment

The reason that scientists want to get that poo back is because of what might be growing in those bags. One-tenth of poo is made up of **MICROBES.** It would be very interesting to find out what has happened to those microbes after more than fifty years on the moon.

Poo Come HOME

What can the poo on the moon tell us? The moon is very different from Earth. Almost all of the things that make life possible on Earth are missing on the moon. If the microbes in the poo have survived on the moon, then it might be possible that there is life elsewhere in space.

Right now, scientists are trying to find out if there is (or ever was) life on the planet Mars.

SANDWICHES in Space

Eating in space is almost as difficult as using the toilet in space, and it is twice as dangerous. In space, loose cookie crumbs could blind an astronaut or get into the machinery and cause serious problems. Bread was at the top of the list of banned food in space. That means no sandwiches, pizza, toast, or bagels.

Scientists have invented crumb-free bread, making some astronauts very happy.

Food BANNED from Space

Some of the other foods banned from space include these:

FIZZY DRINKS
Without gravity, fizzy drinks don't have any bubbles, and warm, flat, fizzy drinks are disgusting.

FRESH MILK
You will only find **DEHYDRATED MILK** in space, because it doesn't need to go in the fridge.

Roast BEEF in Space

In 1965, NASA launched a mission to test a few things. One of those things was space food. One astronaut thought it would be funny to SMUGGLE a roast beef sandwich into space. It wasn't. Astronauts John Young and Gus Grissom were the first and last people to eat a sandwich in space. After just a few bites of the sandwich, they noticed that it was breaking up and spreading lots of dangerous crumbs everywhere.

NASA tightened up their security after this, and no other banned food was ever smuggled into space again—that we know of.

The Worst BURPS

When astronauts burp in space, they actually throw up. This is because gravity keeps down the solids and liquid in your stomach when you burp on Earth, so only gas comes out.

A burp in space is sometimes called a wet burp. Ew.

SALT & PEPPER
These have been replaced with liquid versions—imagine putting liquid salt on your chips.

Early space food had a coating of GELATIN. This would help the food stick together and not break apart into small bits in MICROGRAVITY.

Moon

GOLF

The Golf Swing That Was Out of This WORLD

In 1971, Alan Shepard hit two golf balls on the moon. It is believed that the second ball he hit traveled somewhere between 590 and 1,200 feet (180 and 366 m). The longest golf drive (that is just what golfers call a hit) on Earth was 1,542 feet (470 m).

The SCIENCE behind the Swing

Alan Shepard's swing should have beaten the record on Earth by miles. Literally by miles. But it didn't—so why not? On the moon there is no air or wind to cause **WIND RESISTANCE**, and the gravity on Earth is six times stronger than it is on the moon. The ball should travel a lot farther without these **FORCES** working on it. Right?

So, what went wrong? Well, Shepard had to wear a bulky, uncomfortable space suit, which made it much harder to hit the golf ball. Shepard also had to take the swing with one hand.

Look how bulky that suit is!

What If a GOOD Golfer Hit the Ball on the Moon?

A good golfer can hit the ball around 885 feet (270 m). When a golfer swings, the club travels around 87 miles (140 km) per hour and sends the ball through the air at around 180 miles (290 km) per hour. This kind of power on the moon would send the ball around 12,953 feet (3,948 m).

It would take you around forty-eight minutes to walk 12,953 feet.

Space CAR

There is a red electric sports car in space right now. It is floating around aimlessly. The company that sent the car into space is called Space X. The man in charge of Space X, Elon Musk, says that the car radio is playing the David Bowie song "Space Oddity" at full blast. But no one can hear it.

In Space, No One Can Hear Your RADIO

Space is a vacuum. That means that it has no air in it. Sound cannot travel without air. We hear sound because it causes **VIBRATIONS** in the air. When those vibrations reach our ears, we hear the sound. So, a stereo playing in space is just very silly.

But . . . WHY?

There was a good reason for sending this car into space. Space X needed to test their rocket, the Space X Falcon Heavy rocket, and see if it could release a payload. A payload is just something that is carried into space. In the future, the rocket would probably carry a **SATELLITE**. However, this time it was an electric sports car.

The Longest Car JOURNEY in History

The car even has a driver, and his name is Starman. Don't worry—he is just a dummy in a space suit.

Believe it or not, some experts think that the sports car will one day return to Earth. After going around the sun many times, it is believed that the car will crash into either Venus or Earth in the next few tens of millions of years.

Will the radio still be playing when it comes back?

Don't CRY Over Spilled Astro-Milk

Imagine that you are an astronaut floating around your spacecraft. Suddenly, you float right into the ceiling and bump your head. Whatever you do, don't cry! I mean it—you won't like what happens if you cry.

Things that you take for granted on Earth can be big problems when you are in space.

There is a good reason not to cry in space, and that reason is gravity. When we cry on Earth, the tears leave our eyes and roll down our cheeks. It is gravity that makes the tears go down. And without gravity, we get tear goggles.

When astronauts cry, the tears build up into big, wobbly balls of liquid around the eye. Usually, liquids take the shape of the container that they are in, such as water in a cup or a pond. However, in space, liquid just forms spheres.

The big question is how do you hide the fact that you have been crying when big balls of tears are floating around the space station?

Astronaut Chris Hadfield said that space tears actually hurt a bit, but that when the tear balls get big enough, they just float away.

Emergency SPONGE

Astronaut Andrew Feustel was on a **SPACE WALK** when something got in his eye. While on a space walk, astronauts wear full space suits, including a helmet. Feustel couldn't rub his eye, and the tears began to build up into orbs. Then he had a bright idea. Inside the helmet is a small sponge used to plug astronauts' noses. He used it to rub his eyes. Gross but clever.

Boiling TONGUES

Why the BIG Space Suits?

Human beings really don't do well in space. This is why we have to build all these fancy space stations, rockets, and space suits to protect us. Space is a vacuum. It doesn't have any stuff in it—not even air—which means there is almost no pressure. Pressure is the sense of air or water pushing against you.

LESSON Learned

What happens to somebody in space if they don't have a space suit? Would they **EXPAND** until they pop? Would they pass out from a lack of air? NASA worker Jim LeBlanc found out when his suit broke while carrying out some tests in a vacuum chamber.

Hubble, Bubble, BOIL and Trouble

LeBlanc very quickly lost CONSCIOUSNESS. The last thing that he remembered was all the saliva on his tongue bubbling and boiling. He was quickly rescued, but this accident taught NASA an important lesson. In a vacuum, liquids boil at a lower temperature: body temperature!

A boiling tongue would feel unpleasant.

Expanding BUBBLES

The boiling causes bubbles to form in your bodily fluids, such as your blood. This is called ebullism. It is believed that your body could double in size because of the bubbles in your blood. Even worse, the gas in your blood would expand in the vacuum of space, making you even bigger!

You would be like a big balloon full of boiling blood!

Raindrops Keep SMASHING on My Head

Have you ever been caught in the rain without an umbrella? It can be very annoying to get wet. But just imagine if each one of those raindrops were a diamond. Well, on Jupiter, that is exactly what happens. When it rains, it rains diamonds.

Diamonds are one of the hardest things on Earth.

Jupiter is a gas giant. Gas giants are mostly made of gas, but Jupiter also has a liquid core that is hotter than the surface of the sun. Don't worry—it is very unlikely that you will ever get caught in a diamond rainstorm on Jupiter. It is too hot for starters, and there is nothing solid to stand on.

Why Does It Rain DIAMONDS?

Jupiter is covered in powerful lightning storms that rage across its surface. These lightning storms turn methane (a type of gas we have here on Earth—you can find it in farts) into soot. As the soot falls, it first hardens into graphite and then into diamonds as it gets closer to the hot core.

The diamonds on Jupiter melt when they reach the **MOLTEN** liquid core of the planet.

No scientists have been to Jupiter. They have just figured out what it might be like by using science. If they did go to Jupiter, they would be smashed to smithereens by the diamonds. At the same time, they would fall into the molten liquid core and suffocate from not having any air to breathe.

One Small Step for Man, Eight Tiny Steps for a SPIDER

So, to go into space you need to train for years to be an astronaut, right? Wrong! If you are a spider, cat, dog, or even a tortoise, you might be randomly picked to be sent into space. In fact, a surprising number of animals have been to space. Here are just some of them.

1961:
a chimpanzee

1963:
Félicette,
a French cat

1968:
a Russian
tortoise

1973:
garden
spiders

HAM

Ham is one of the most famous nonhuman astronauts. Ham was a chimpanzee who was fired into space for a sixteen-minute flight. His safe return proved to NASA that the flight would be safe for the human astronauts, Alan Shepard and Gus Grissom.

The Cow JUMPED Over the Moon

The Moon WALK

When Neil Armstrong and Buzz Aldrin walked on the moon in 1969, the world saw how strange it was to walk in low gravity. But what happens when you jump in low gravity? Earth's gravity is six times stronger than the moon's, which means that on the moon you could jump around six times higher!

Moon JUMP

The average jump height for humans on Earth is between 12 and 16 inches (30 and 41 cm). Not a bad effort, humans. But there is an animal on Earth that puts our human jumps to shame. The red kangaroo can jump up to 72 inches (183 cm)—that's as high as the average adult human male stands.

A human who could jump 16 inches (41 cm) on Earth would be able to jump 7 feet 10 inches (2.4 m) high on the moon. Wow, good work, human! However, a red kangaroo would be able to jump 36 feet (11 m). That is higher than two and a half double-decker buses stacked on top of each other. Don't worry, though—even a kangaroo couldn't jump off the moon.

Can the Cow REALLY Jump Over the Moon?

No! But cows can jump higher than you think. The highest jump recorded by a cow was around 72 inches (183 cm). That's as high as a kangaroo's jump! However, the cow must have had a good running start. It is unlikely that a cow would land on its feet from a 36-foot (11 m) jump on the moon.

Space Is WEIRD!

PROS

Space is pretty cool! You can jump very high on the moon, and you are taller than usual. Not many people have been to space, and there is still lots to learn.

CONS

Space is weird! Jumping on the moon looks like fun, but it would be no fun to damage your space suit and have your blood boil. Then you would cry about it and the tears would be stuck in your eyes. There are no sandwiches, and let's not forget about going to the toilet. Toilets are much neater with gravity on your side.

GLOSSARY

Apollo 10 mission a 1969 space mission to fly around the moon

consciousness an alert state in which someone is awake and aware of what is happening

dehydrated milk a dry powder that can be turned into milk by adding water

deprived to be without something

expand to get bigger

forces invisible energies that push, pull, or move objects

gelatin a colorless and tasteless ingredient that is added to food to make it thick and gloopy

gravity the force that pulls everything down toward the center of large objects in space

lasers machines that make very strong beams of light

microbes tiny living things that can only be seen with a microscope; especially bacteria, viruses, and fungi

microgravity very weak gravity

molten melted by extreme heat

NASA the National Aeronautics and Space Administration, the US space agency

percent one part in every one hundred

redistributed to be spread out differently

satellite a machine in space that travels around planets taking photographs and collecting and sending information

smuggle to secretly bring something in or out of a place in which it is not allowed or not allowed to be removed from

space agencies organizations that research and explore space

space walk a mission in which astronauts wear space suits to go outside the spacecraft

vibrations movements going in any direction very quickly

wind resistance the frictional force air creates against a moving object

INDEX

PHOTO CREDITS: Images are courtesy of Shutterstock.com. With thanks to Getty Images, Thinkstock Photo, and iStockphoto.

Cover: bosotochka, recurring images: bosotochka (header font), Sonechko57 (blobs), Aphelleon (photo space background), PremiumArt (vector space background), Amy Li (additional illustrations). P2–3 – Dima Zel, p4–5 –wavebreakmedia, Olian_RosaYuganov Konstantin, M.Aurelius, Viktor1, Mooi Design, Andrei Verner, Nadya_Art p6–7 – 3Dsculptor,iurii, BlurryMe, victoruler, p8–9 – Professional Bat, gst, Yindee, Yuganov Konstantin, p10–11 – Igor Kardasov, Sudowoodo, Top Vector Studio, yusufdemirci, Vectorpocket, p12–13 –Castleski, sudowoodo, Hvsht, Professional Bat, Roman Samborskyi, KiattisakCh, p14–15 – PremiumArt, yatate, farbled, Rvector, VikiVector, Elena Elisseeva, Hurst Photo, kzww, p16–17 – Yindee, taffpixture, Zakharchenko Anna, solar22, p18–19 – oleskalashnik, Dmitriy Nikiforov, ForeverLee, YMZK-Photo, Real Illusion, SergeyDV, p20–21 – Lorelyn Medina, Samantha Cristoforetti – wiki com, Lia Koltyrina, p22–23 – Andrey Armyagov, Igdeeva Alena, NotionPic, oljalja, WAYHOME studio, p24–25 – blocberry, CHAjAMP, Ilya Bolotovmaryna rodyukova, Nerthuz, p26–27 – Maria Skrigan, NASA (wiki commons), AlexHliv, HappyPictures, Nadezda Barkova, natchapohn, NotionPic, Sunnydream, wet nose, yusufdemirci, p28–29 – maryna rodyukova, ashva, Professional Bat, mr.barghest, Merlin74.